Pre-Lab I

Experimental Organic Chemistry

A MINISCALE APPROACH

Royston M. Roberts
John C. Gilbert
Stephen F. Martin

University of Texas, Austin

Saunders Golden Sunburst Series
Saunders College Publishing
Harcourt Brace College Publishers

Fort Worth Philadelphia San Diego New York Orlando
San Antonio Austin Toronto Montreal London Sydney Tokyo

Copyright ©1994 by Saunders College Publishing

All rights reserved. No part of this publication may be reproduced or transmitted in any form or by any means, electronic or mechanical, including photocopy, recording, or any information storage and retrieval system, without permission in writing from the publisher.

Although for mechanical reasons all pages of this publication are perforated, only those pages imprinted with a Harcourt Brace & Company copyright notice are intended for removal.

Requests for permission to make copies of any part of the work should be mailed to: Permissions Department, Harcourt Brace & Company, 8th Floor, Orlando, Florida 32887

Printed in the United States of America.

Roberts/Gilbert/Martin: Pre-Lab Exercises for EXPERIMENTAL ORGANIC CHEMISTRY: A MINISCALE APPROACH

ISBN 0-03-097284-1

45 021 98765432

Pre-Lab Exercises for RECRYSTALLIZATION
(Section 3.2)

NAME (print): _____ DATE: _____

INSTRUCTOR: _____ LABORATORY SECTION: _____

1. By marking yes (Y) or no (N), specify which of the following criteria are met by a good solvent for a recrystallization.

   Y   N
   ___ ___ a. The solutes are soluble in the cold solvent.
   ___ ___ b. The solvent does not react chemically with the solutes.
   ___ ___ c. The solvent is polar rather than nonpolar.
   ___ ___ d. The boiling point of the solvent is above 100 °C.
   ___ ___ e. The boiling point of the solvent ideally is below the melting point of the solute being purified.

2. Indicate which of the two filtration techniques, gravity (G) or vacuum (V), is the more suitable for each of the following operations.

   G   V
   ___ ___ a. Hot filtration.
   ___ ___ b. Removing decolorizing carbon.
   ___ ___ c. Isolating recrystallized solute from solution.

3. Why is a flame *not* used to heat a solution in hexane or diethyl ether during a recrystallization?

4. Why should the size of crystals obtained in a recrystallization be neither too large nor too small?

5. What is the process of seeding, as it applies to recrystallization? What purpose does it serve?

6. How is the purity of a recrystallized solid assessed?

7. Give the criterion applied in this experimental procedure to classify a solute as being "soluble" in a particular solvent.

8. Why should decolorizing carbon *not* be added to a solvent that is at or near its boiling point?

9. Review the functional groups present in resorcinol, benzoic acid, naphthalene, and acetanilide. Predict whether these molecules are expected to be polar (P) or nonpolar (NP).
   Resorcinol _____   Benzoic acid _____   Naphthalene _____   Acetanilide _____

10. Why is it important to
    a. break (terminate) the vacuum before turning off the water aspirator pump when employing the equipment for vacuum filtration shown in Figure 2.28?

    b. avoid the inhalation of vapors of organic solvents?

    c. know the position and procedure of operation of the nearest fire extinguisher when using methanol, 95% ethanol, or 2-propanol as a crystallization solvent?

    d. use a *fluted* filter paper for hot filtration?

11. Values of the oral LD_{50} (mg/kg) in rats of resorcinol, benzoic acid, naphthalene, and acetanilide are _____, _____, _____, and _____, respectively.

Pre-Lab Exercises for MELTING POINTS
(Section 3.3)

NAME (print): _____ DATE: _____

INSTRUCTOR: _____ LABORATORY SECTION: _____

1. List a convenient source of data concerning the physical constants and properties of organic compounds.

2. Indicate which of the following statements is true (T) and which false (F) by putting a check mark in the appropriate space.

   T   F
   ____ ____ a. An impurity raises the melting point of an organic compound.
   ____ ____ b. A eutectic mixture has a sharp melting point, just as does a pure compound.
   ____ ____ c. If the rate of heating of the oil bath used in a melting-point determination is too high, the melting point that results will likely be too low.
   ____ ____ d. The sample should not be packed tightly into a capillary melting-point tube.
   ____ ____ e. A heating bath containing mineral oil should not be used to determine the melting points of solids melting above 200 °C.

3. On the figure below, sketch the location of the capillary melting-point tube and the sample contained in it relative to the bulb of the thermometer.

   insert thermometer picture

4. What is the approximate rate at which the temperature of the heating bath should be increasing at the time the sample undergoes melting?

5. What is the preferred technique for accurately determining the melting point of an unknown compound in a minimum length of time?

6. How does measuring a mixture melting point help in determining the possible identity of two solid samples?

7. Briefly describe the technique for packing a capillary melting-point tube.

8. Why is it important to calibrate a thermometer with a set of standards having a *range* of melting points?

9. What toxic fumes are evolved by burning mineral oil?

Pre-Lab Exercises for MICRO BOILING POINTS (Section 4.1)

NAME (print): _____ DATE: _____

INSTRUCTOR: _____ LABORATORY SECTION: _____

Questions 1, 2, and 4 may be answered by marking yes (Y) or no (N) for each part.

1. The addition of a nonvolatile solute to a volatile liquid
   Y   N
   ___ ___ a. has no effect on the boiling point of the volatile liquid.
   ___ ___ b. lowers the boiling point of the volatile liquid.
   ___ ___ c. raises the boiling point of the volatile liquid.

2. The boiling point of a pure liquid
   Y   N
   ___ ___ a. is the same in Denver, Colorado (elevation 5280 ft) as it is in San Francisco, California.
   ___ ___ b. is lower in Denver than in San Francisco.
   ___ ___ c. is usually found to be almost the same as the reported "standard boiling temperature" in San Francisco.

3. Explain your answer to exercise 2.

4. The boiling point, as determined in the micro boiling point apparatus, is the temperature at the time
   Y   N
   ___ ___ a. bubbles first emerge slowly from the inverted capillary tube.
   ___ ___ b. bubbles begin to emerge rapidly from the inverted capillary tube.
   ___ ___ c. the liquid begins to re-enter and rise in the inverted capillary tube.

5. What are the bubbles that emerge slowly from the inverted capillary tube before the boiling temperature is reached, and why does this occur?

6. Define a *closed system* as it applies to a Thiele tube and the liquids it contains.

7. Why should a closed system, as defined in exercise 6, *not* be heated unless special apparatus is available?

8. Why is the micro boiling point apparatus used in this procedure *not* a closed system?

9. Why is mineral oil an inappropriate heating fluid for determining the boiling points of samples that exceed 200 °C?

Pre-Lab Exercises for SIMPLE AND FRACTIONAL DISTILLATION (Section 4.2)

NAME (print): _____ DATE: _____

INSTRUCTOR: _____ LABORATORY SECTION: _____

1. Indicate which of the two distillation techniques, simple (S) or fractional (F), would be more suitable for the following operations by putting a check mark in the appropriate space.

    S   F
    ___ ___ a. preparing drinking water from sea water.
    ___ ___ b. removing diethyl ether, bp 35 °C (760 torr), from a solution containing p-dichlorobenzene, bp 174 °C (760 torr).
    ___ ___ c. separating benzene, bp 80 °C (760 torr), from toluene, bp 111 °C (760 torr).

2. What is the purpose of the boiling chips or the magnetic stirring bar placed in a stillpot?

3. How does the composition of the liquid at the top of a fractional distillation column compare with the composition of the liquid at the bottom of a column? (Answer in terms of the relative amounts of lower-boiling and higher-boiling components.)

4. Two fractionating columns are each 40 cm in length. Column A has HETP = 2 cm. and column B has HETP = 20 cm. Which column, A or B, would be more appropriate to separate a binary mixture in which the components differ in boiling point by 10 °C?

    A_____   B_____

5. Define the term, *reflux ratio*.

6. Why is it important to align the fractionating column as nearly vertical as possible?

7. On the figure below, sketch the correct location of the thermometer bulb.

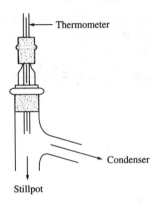

8. With respect to the condenser used in an apparatus for simple or fractional distillation, why should the lower rather than the upper nipple be used for the water inlet?

9. The flash point (°C) of ethyl acetate is _____; that of n-butyl acetate is _____.

10. The concentration, in ppm, at which human eye irritation is produced by n-butyl acetate is _____, and by ethyl acetate _____.

Pre-Lab Exercises for STEAM DISTILLATION OF CITRAL (Section 4.2.7)

NAME (print): _____ DATE: _____

INSTRUCTOR: _____ LABORATORY SECTION: _____

1. Indicate which of the following statements is true (T) or false (F) by placing a check mark in the appropriate space.
   Steam distillation would be the procedure of choice for separating mixtures of

   T   F
   ____ ____ a. methanol, bp 65 °C (760 torr), and water (methanol is completely miscible with water).
   ____ ____ b. p-dichlorobenzene, bp 174 °C (760 torr), and water; p-dichlorobenzene is not soluble in water.
   ____ ____ c. Ethylene glycol ($HOCH_2CH_2OH$), bp 196 °C (760 torr), and water; the glycol is miscible with water in all proportions.

2. Explain your answers to each part of question 1.

   a.

   b.

   c.

3. The boiling point during a steam distillation is always less than 100 °C.

   T_____   F_____

4. Explain your answer to question 3.

5. In what way are the advantages of steam distillation and vacuum distillation similar?

6. What kind of mixture may better be separated by steam distillation than by vacuum distillation?

7. Write structural formulas for the two diastereomers (geometrical isomers) of citral.

8. Based on a consideration of the nature of the groups present in the diastereomers that comprise citral, why is citral insoluble in water?

9. Why is a flame *not* used to assist in the removal of diethyl ether from citral?

10. Why should ethereal solutions *not* be stored in your laboratory locker from one period to the next?

Pre-Lab Exercises for ISOLATION OF TRIMYRISTIN FROM NUTMEG (Section 5.3)

NAME (print): _____ DATE: _____
INSTRUCTOR: _____ LABORATORY SECTION: _____

1. Name the two most common methods for isolating a desired organic substance from a plant source.

2. Answer true (T) or false (F) by checking the appropriate space.

    T   F
    ___ ___ a. It is more efficient to perform *three* extractions with 10-mL portions of solvent than to perform *one* extraction with 30 mL of the same solvent.
    ___ ___ b. According to the equation defining the distribution coefficient K, a value of 2 for K means that A is more soluble in solvent S than in solvent S'.

    $$K = \frac{\text{grams of } A \text{ in } S/\text{mL of } S}{\text{grams of } A \text{ in } S'/\text{mL of } S'}$$

    ___ ___ c. Nutmeg contains a complex mixture of organic compounds that are soluble in diethyl ether.
    ___ ___ d. Trimyristin is more soluble in acetone than in diethyl ether.
    ___ ___ e. The functional group present in trimyristin is a carboxylic acid group.
    ___ ___ f. The term *lipid* describes a category of organic compounds that are insoluble in water.
    ___ ___ g. With respect to the technique of liquid-liquid extraction, the term *partitioning* means physical separation of two immiscible liquids by an impermeable membrane.

3. Explain your answer to Exercises 2a, 2e, and 2f.
    a.

    e.

    f.

4. Why is trimyristin *not* optically active?

5. What is wrong with the experimental set-up shown below for extraction of nutmeg?

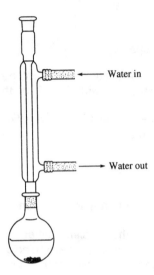

6. On the figure shown in Exercise 5, indicate on the condenser the *upper* limit for the ring of condensate and the points at which clamps should be located.

7. Why is a flame *not* to be used to heat diethyl ether at reflux in this experiment?

8. The flash point (°C) of diethyl ether is _____ ; that of acetone is _____ .

9. The concentration, in ppm, at which human eye irritation is produced by diethyl ether is _____ , and by acetone is _____ .

Pre-Lab Exercises for ACID AND BASE EXTRACTION
(Section 5.4)

NAME (print): _____ DATE: _____

INSTRUCTOR: _____ LABORATORY SECTION: _____

1. Write equations for all chemical reactions that occur in the extraction procedure.

2. Devise a flow chart for purification that summarizes the separations performed in this experiment. Consult Figure 1.2 for an example of such a flow chart.

3. When extracting an aqueous solution with an organic solvent, if you are uncertain as to which layer in the separatory funnel is aqueous, how could you settle the issue?

4. Which layer, upper (U) or lower (L), will each of the following solvents usually form when used to extract a dilute aqueous solution: diethyl ether _____, dichloromethane _____, chloroform _____, hexane _____?

5. With respect to the neutralization step in the procedure, calculate the amount of base required to neutralize the acidic extract and the amount of acid needed to neutralize the basic extract.
   a. Base required:

   b. Acid required:

6. Indicate whether each of the following statements is true (T) or false (F) by checking the appropriate space.
   _____ _____ a. Benzoic acid and *p*-nitroaniline form water-soluble salts, whereas anthracene does not.
   _____ _____ b. Carboxylic acids and phenols containing six or more carbon atoms per molecule are more soluble in dichloromethane than in water.
   _____ _____ c. Carboxylic acids and phenols containing six or more carbon atoms per molecule are more soluble in 3 *M* sodium hydroxide than in dichloromethane.
   _____ _____ d. Anthracene is more soluble in dichloromethane than is sodium benzoate.

7. What toxic fumes are produced upon combustion of dichloromethane and chloroform?

Pre-Lab Exercises for COLUMN CHROMATOGRAPHY (Section 6.2)

NAME (print): _____ DATE: _____
INSTRUCTOR: _____ LABORATORY SECTION: _____

1. What difficulty may result if
   a. a chromatographic column is not placed in a vertical position?

   b. the liquid level of the eluent is allowed to drop below the top of the column?

2. Petroleum ether followed by dichloromethane is used to separate fluorene and fluorenone. Why would reversing the order in which these solvents are used be unwise?

3. What is the best experimental procedure to use in choosing an eluting solvent for column chromatography?

4. Define:
   a. eluent

   b. eluate

    c. adsorption

5. Which type of alumina, "acidic" or "basic," would provide for the better separation of acids?

6. How can the adsorptivity (activity) of alumina be varied?

7. Why should a mixture to be separated be introduced onto a column in a minimum amount of solvent?

8. Why should no flames be allowed on the lab bench when performing this experiment?

9. Underline the media that are appropriate for extinguishing fires involving fluorene or fluorenone: Water   Carbon dioxide   Chemical powder   Foam

10. With what general class of chemical reagents are fluorene and fluorenone incompatible?

Pre-Lab Exercises for THIN-LAYER CHROMATOGRAPHY (Section 6.3)

NAME (print): _____ DATE: _____

INSTRUCTOR: _____ LABORATORY SECTION: _____

1. Describe three ways in which colorless compounds can be located on a TLC plate.

2. Check TLC (thin-layer chromatography) or CC (column chromatography) as the more appropriate answer to the following questions or statements.
   a. TLC _____ CC _____ is a quicker procedure for separating components of a mixture.
   b. In TLC _____ CC _____ the solvent front moves downward.
   c. TLC _____ CC _____ is better for separating a 5-gram mixture of components.
   d. TLC _____ CC _____ is better for separating a mixture of volatile compounds.

3. Why should a TLC plate be removed from the solvent *before* the solvent front reaches the top of the plate?

4. What is the purpose of shaking the petroleum ether-ethanol extract of the leaves with water in a separatory funnel?

5. The separating power, or activity, of a TLC plate is increased by heating the plate in an oven at 100 °C. Why? (*Hint:* See section 6.2.1)

6. Why should the developing chamber for a TLC plate *not* be open to the atmosphere?

7. Which of the following diagrams illustrate(s) an *improper* way of spotting a TLC plate? Tell what is wrong in each case.

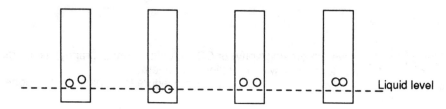

8. What toxic fumes are evolved in fires involving azobenzene?

9. The oral toxicity (g/kg) of azobenzene in rats is _____.

Pre-Lab Exercises for GAS-LIQUID CHROMATOGRAPHY
(Section 6.4)

NAME (print): _____ DATE: _____

INSTRUCTOR: _____ LABORATORY SECTION: _____

1. Explain how *liquids* can be analyzed by gas chromatography.

2. Briefly define or describe the function, in gas chromatography, of the
    a. carrier gas

    b. stationary liquid phase

    c. solid support

3. Which column material should be chosen to separate a mixture of
                                        Carbowax 20M   SE 30
    a. alcohols   _____   _____
    b. aromatic hydrocarbons   _____   _____

4. If ethyl acetate and *n*-butyl acetate are analyzed by gas chromatography, which of these esters will generally produce a peak with the *shorter* retention time?

5. How could you confirm your answer to question 4 by experiment?

6. What operating variables determine retention time?

7. The flash point (°C) of ethylbenzene is _____, of toluene is _____, and of isopropylbenzene is _____.

8. Underline the media that are appropriate for extinguishing fires involving the hydrocarbons listed in exercise 7:   Water   Carbon dioxide   Chemical powder   Foam

**Pre-Lab Exercises for SEPARATION OF DIASTEREOMERIC
1,2-CYCLOHEXANEDIOL (SECTION 7.2)**

NAME (print): _____ DATE: _____

INSTRUCTOR: _____ LABORATORY SECTION: _____

Answer true (T) or false (F) by checking T or F in Exercises 1-4:

T   F

____ ____ 1. *trans*-1,2-Cyclohexanediol is a *meso* form.

____ ____ 2. *cis*- and *trans*-1,2-Cyclohexanediols are enantiomers.

____ ____ 3. *cis*- and *trans*-1,2-Cyclohexanediols cannot be separated by fractional crystallization.

____ ____ 4. (+)-*trans*-Cyclohexanediol and (-)-*trans*-cyclohexanediol have different chromatographic adsorption properties.

5. Why should a TLC plate be removed from the solvent before the solvent front reaches the top of the plate?

6. Why should the developing chamber for a TLC plate not be open to the atmosphere?

7. What physical properties could be used to distinguish *cis*- from *trans*-1,2-cyclohexanediol?

8. Why might water *not* be an appropriate extinguishing medium for burning petroleum ether?

9. Identify which of the following diagrams illustrates an *improperly* spotted TLC plate and explain what is wrong in each such case.

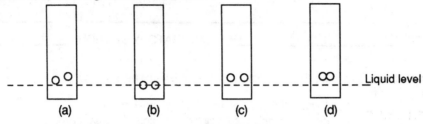

10. The flashpoints (°C) for petroleum ether, for acetone, and for 2-propanol are _____, _____, and _____, respectively.

Pre-Lab Exercises for **ISOMERIZATION OF DIMETHYL MALEATE TO DIMETHYL FUMARATE** (SECTION 7.3)

NAME (print): _____ DATE: _____
INSTRUCTOR: _____ LABORATORY SECTION: _____

1. Write structural formulas for dimethyl maleate and dimethyl fumarate.

2. Check the correct answer: Dimethyl maleate and dimethyl fumarate are
   a. enantiomers _____, (b) diastereomers _____, (c) constitutional isomers _____.

3. Which of the following is *least* likely to be responsible for the isomerization of dimethyl maleate? (a) Br^+, (b) Br^-, (c) $Br^\cdot$.

4. Which ester
   a. has the higher melting point,   dimethyl maleate _____ or dimethyl fumarate _____?
   b. has the higher boiling point,   dimethyl maleate _____ or dimethyl fumarate _____?
   c. is more soluble in CCl_4,   dimethyl maleate _____ or dimethyl fumarate _____?

5. Why should glassware containing residues of bromine *not* be rinsed with acetone?

6. State the recommended procedure in this experiment for destroying residual bromine and write the equation for the reaction that occurs.

7. What is the vapor pressure of bromine at 20 °C?

8. Why should inhalation of the vapors of bromine be avoided?

9. Provide the flashpoint (°C) of carbon tetrachloride _____ ; of ethanol _____ .

Pre-Lab Exercises for ISOLATION OF THE ENANTIOMERS OF CARVONE FROM ESSENTIAL OILS (Section 7.5)

NAME (print): _____ DATE: _____

INSTRUCTOR: _____ LABORATORY SECTION: _____

1. Why is vacuum distillation used to separate the carvone isomers from the other compounds present in spearmint and caraway seed oils?

2. Why is an electrically heated oil bath better than a heating mantle to heat a stillpot in a distillation?

3. Why is a 50-mL stillpot better than a 100-mL stillpot for distillation of 30 mL of a liquid?

4. Why is heating discontinued while changing receivers in a vacuum distillation?

5. Why should vacuum be established before heating of the stillpot is begun?

6. *Underline* any of the following physical properties of the enantiomeric carvones that would be expected to be the same: bp, solubility in acetone, R_f value in TLC, odor, retention time in GLC, rotation of plane-polarized light.

7. Why should heavy-walled rather than thin-walled rubber tubing be used to connect the apparatus to the vacuum source?

8. Why should the stillpot and receivers be weighed prior to assembly of the distillation apparatus?

9. The flashpoints (°C) of ethanol, ethyl acetate, and carvone are _____, _____, and _____, respectively.

10. What can occur if 2,4-dinitrophenylhydrazine is absorbed into the body?

**Pre-Lab Exercises for RESOLUTION OF RACEMIC 1-PHENYLETHANAMINE
(Section 7.6)**

NAME (print): _____ DATE: _____
INSTRUCTOR: _____ LABORATORY SECTION: _____

1. Methanol is used as the solvent for the separation of the diastereomeric forms of 1-phenylethanamine hydrogen tartrate. Briefly explain why this solvent is better for the separation than
   a. petroleum ether (bp 60–80 °C).

   b. water.

2. Why is the solution of 1-phenylethanamine hydrogen tartrate not cooled in *ice water* before the crystals are collected?

3. Why cannot racemic tartaric acid be used to resolve 1-phenylethanamine?

4. Why cannot *meso* tartaric acid be used to resolve 1-phenylethanamine?

5. What is the crystalline form of the amine hydrogen tartrate to be isolated in this experiment?

6. *Underline* any of the following physical properties of the enantiomers of 1-phenylethanamine that would be expected to be the same: bp, solubility in hexane, odor, retention time in GLC, rotation of plane-polarized light.

7. The flashpoints (°C) of diethyl ether, methanol, and 1-phenylethanamine are _____, _____, and _____, respectively.

8. What toxicology data, if any, are provided in the MSDS sheets for tartaric acid?

Pre-Lab Exercises for CHLORINATION USING SULFURYL CHLORIDE (Section 9.2)

NAME (print): _____ DATE: _____

INSTRUCTOR: _____ LABORATORY SECTION: _____

1. Write the structure of the substance used in these experiments to initiate free radical chain chlorination, and give the equation for its thermal decomposition.

2. What is the molar ratio of initiator to sulfuryl chloride used in these procedures? Why is so little initiator needed relative to the amount of chlorinating agent used?

3. Write the equation for the reaction of sulfuryl chloride with water and explain why water would be an inappropriate medium for extinguishing a fire involving this chemical..

4. Why must a "gas trap" be used in these experimental procedures, and why is the trap equipped so that air pulled into it must first pass through calcium chloride?

5. What is the theoretical amount of weight that should be lost by the reaction mixture in the particular procedure you are to perform? Show your calculation.

6. Why should the separatory funnel be vented frequently when the reaction mixture is shaken with aqueous sodium carbonate? Write an equation for any chemical reaction that is responsible for the need for venting.

7. What type of distillation apparatus, fractional or simple, is used for isolation of the chlorinated product(s)? Why is this type selected?

8. Why are pieces of glass or of stainless steel "sponge" rather than copper "sponge" recommended as packing material for the distillation column?

9. Put a checkmark beside each of the following materials that evolves toxic fumes upon heating or burning: Cyclohexane _____, sulfuryl chloride _____, sodium chloride _____, 1-chlorobutane _____, sodium sulfate _____.

10. The oral LDL_o (mg/kg) in humans of bromine is _____.

Pre-Lab Exercises for BROMINATION: SELECTIVITY OF HYDROGEN ATOM ABSTRACTION AS A FUNCTION OF STRUCTURE (Section 9.3)

NAME (print): _____ DATE: _____

INSTRUCTOR: _____ LABORATORY SECTION: _____

1. Draw a structure containing the specified type of hydrogen atom and circle the atom.
   a. 1° aliphatic hydrogen atom.

   b. 3° benzylic hydrogen atom.

   c. vinylic hydrogen atom.

2. What experimental criterion is to be used to measure the rates of bromination of the hydrocarbons that are used in this experiment?

3. Underline the proper response in the following:
   a. The results of this experiment allow determination of the (relative, absolute) rates of bromination of a series of hydrocarbons.
   b. The mechanism of the bromination is classified as a(n) (electrophilic addition, nucleophilic substitution, free-radical substitution) process.
   c. Molecular bromine is a (solid, liquid, gas) at room temperature and is (corrosive, non-corrosive) to the skin.

4. Why should apparatus containing residues of bromine *not* be rinsed with acetone? How can such residues be chemically removed?

5. Calculate the average ratio of Br_2 to hydrocarbon used in this experiment. Assume a density of 0.8 and a molecular weight of 100 for the calculation. Which reagent is used in excess and why?

6. Why should the temperature of the water bath used to heat the reaction mixtures not exceed about 50 °C?

7. Why is the bromine to be added to each test tube in this experiment measured out as a solution of bromine in carbon tetrachloride rather than as pure bromine?

8. The oral LDL_o (mg/kg) in humans of bromine is _____ and of toluene is _____.

9. The flash point (°C) of toluene is _____; that of methylcyclohexane is _____.

10. Underline the media that are appropriate for extinguishing fires involving toluene, *tert*-butylbenzene and ethylbenzene: Water   Carbon dioxide   Chemical powder   Foam

Pre-Lab Exercises for DEHYDROHALOGENATION OF 2-CHLORO-2-METHYLBUTANE (Section 10.2)

NAME (print): _____ DATE: _____

INSTRUCTOR: _____ LABORATORY SECTION: _____

1. Calculate the molar ratio of base to alkyl chloride used in these experiments (it is the same in each part). Show your calculation. Given the result, what is the "limiting reagent."

2. Why is a Hempel column rather than a regular condenser used during the period of reflux? Why is the column filled with a packing material during this stage of the procedure?

3. Write the structure of the sterically most hindered base used in these eliminations.

4. Will the hold-up of the Hempel column when being used as a fractionating column be greater when packed or unpacked? Why?

5. Why is the vacuum adapter of the apparatus fitted with a calcium chloride drying tube throughout the course of the reaction and the distillation?

6. Why is the receiving flask cooled in an ice-water bath throughout the reaction and distillation?

7. Write equations for the chemical reactions that you will use to demonstrate the presence of alkenes in your distilled product.

8. What is an appropriate extinguishing medium for fires involving sodium methoxide or potassium *tert*-butoxide?

9. Why is water inappropriate for extinguishing a fire involving sodium metal?

10. The flashpoints (°C) of 2-methyl-1-butene and 2-methyl-2-butene are _____ and _____, respectively.

11. The oral LDL_o (mg/kg) in humans of bromine is _____.

Pre-Lab Exercises for DEHYDRATION OF 4-METHYL-2-PENTANOL AND CYCLOHEXANOL (Section 10.3)

NAME (print): _____ DATE:_____

INSTRUCTOR: _____ LABORATORY SECTION: _____

1. What is the function of the acid catalyst in promoting the dehydration of alcohols?

2. Why would concentrated hydrochloric acid be an inappropriate catalyst for the dehydration of alcohols?

3. Why is the formation of substitution products involving displacement of water by attack of bisulfate or of dihydrogen phosphate upon a protonated alcohol *not* a reaction of concern in these procedures?

4. What is the maximum head temperature that should be observed during the dehydration of the alcohols used in these procedures? Why is there this limit?

5. Write equations for the chemical reaction(s) that you will use to demonstrate the presence of alkene(s) in your distilled product.

6. Why is it important to dry the crude alkene(s) prior to the final distillation?

7. How are any acidic materials removed from the crude alkene(s)?

8. What is the "limiting reagent" in the dehydration procedures?

9. Underline the media appropriate for extinguishing fires involving cyclohexene or the isomeric methylpentenes formed by dehydration of the corresponding alcohols: Water   Carbon dioxide   Chemical powder   Foam

10. Which if any of the alkenes formed by dehydration of cyclohexanol or 4-methyl-2-pentanol has a flashpoint above 25 °C?

11. The oral LDL_o (mg/kg) in humans of bromine is _____.

Pre-Lab Exercises for ADDITION OF H-Br TO 1-HEXENE (Section 10.4.1)

NAME (print): _____ DATE: _____

INSTRUCTOR: _____ LABORATORY SECTION: _____

1. Why do the conditions of this experiment favor electrophilic rather than free-radical addition of HBr to 1-hexene?

2. What is the role of the quaternary ammonium salt used in this experiment?

3. When concentrated aqueous HBr is added to 1-hexene, do you expect to observe a homogeneous or a heterogeneous reaction mixture? Why?

4. Why is vigorous stirring or other agitation of the reaction mixture important?

5. Write equations for the chemical reaction(s) that are responsible for the pressure build-up when the crude reaction mixture is washed with aqueous sodium bicarbonate.

6. Write equations for the chemical reactions that could be used to demonstrate that 2- rather than 1-bromohexane has been formed in this experiment.

7. What is the mechanistic basis for each of the reactions you proposed in Exercise 6 that allows differentiation of 1° from 2° alkyl halides?

8. The LC_{50} (ppm) for inhalation of H-Br by rats is _____ over a time period of _____.

9. What toxic fumes are evolved in fires involving methyltrioctylammonium chloride and 1-bromohexane?

Pre-Lab Exercises for HYDRATION OF NORBORNENE
(Section 10.4.2)

NAME (print): _____ DATE:_____
INSTRUCTOR: _____ LABORATORY SECTION: _____

1. Why is it important to add the concentrated sulfuric acid *to* water?

2. What is the function of sulfuric acid in promoting the hydration of norbornene?

3. Why would concentrated hydrochloric acid be an unsuitable replacement for sulfuric acid?

4. What is the purpose of washing the ethereal layer containing the product with saturated sodium bicarbonate and sodium chloride prior to drying the solution with anhydrous sodium sulfate?

5. Write equations for the chemical reaction(s) that are responsible for the pressure build-up when the crude reaction mixture is washed with aqueous sodium bicarbonate.

6. Why is it necessary to seal the capillary tube before determining the melting point of norborneol?

7. What is the purpose of the trap between the water aspirator and the sublimation apparatus?

8. Why is it necessary to remove all of the diethyl ether prior to subliming the product?

9. What is the composition of the solid that might appear in the separatory funnel during the work-up?

10. In rats, the oral LD_{50} (mg/kg) of sulfuric acid is _____.

11. The flashpoint (°C) of norbornene is _____ and its oral LD_{50} (mg/kg) in rats is _____.

Pre-Lab Exercises for HYDROBORATION-OXIDATION OF (+)-α-PINENE (Section 10.4.3)

NAME (print): _____ DATE: _____

INSTRUCTOR: _____ LABORATORY SECTION: _____

1. Calculate the molar ratios of each of the reagents and the (+)-α-pinene. Show your calculations. Given this result, what is the limiting reagent?

2. Why must the solvent and apparatus be dried prior to the reaction of iodine with sodium borohydride?

3. A the end of the reaction, water is added *slowly* to the reaction mixture containing the borane-THF complex. Why?

4. What is the gas that is formed when water is added to the mixture to quench the reaction? Write a balanced equation for the reaction that occurs between water and borane-THF complex.

5. Write out a sequence of balanced equations to show how BH_3 is produced by the reaction of $NaBH_4$ and I_2.

6. Why is it necessary to use *basic* hydrogen peroxide to oxidize the intermediate dialkyl borane?

7. What organic solvent is recommended for recrystallizing the product, and why must water be added to the warm solution of product prior to cooling?

8. Underline the media appropriate for extinguishing fires involving sodium borohydride: Water   Carbon dioxide   Chemical powder   Foam

9. In rats, the oral LD_{50} (mg/kg) of tetrahydrofuran is _____.

10. In humans, the oral LDL_o (mg/kg) of iodine is _____.

11. List three symptoms associated with exposure to I_2.

Pre-Lab Exercises for HYDRATION OF 2-METHYL-3-BUTYN-2-OL (Chapter 11)

NAME (print): _____ DATE: _____

INSTRUCTOR: _____ LABORATORY SECTION: _____

1. Write the structural formula for the product of the reaction of 2-methyl-3-butyn-2-ol with bromine in carbon tetrachloride.

2. Write a balanced equation for the reaction of silver ammonia complex with 2-methyl-3-butyn-2-ol.

3. Why is the silver salt of the alkyne not allowed to dry on the filter?

4. Write an equation for the reaction of dilute hydrochloric acid with the silver salt of the alkyne.

5. Put a checkmark beside the reported chronic effects, if any, of silver nitrate: carcinogen _____, mutagen _____, tumorigen _____.

6. The LHL_o (ppm) for inhalation of ammonium hydroxide by humans is _____ over a time period of _____; the oral LD_{50} (mg/kg) in rats of 2-methyl-3-butyn-2-ol is _____.

7. Circle the media, if any, that are *not* appropriate for extinguishing fires involving 2-methyl-3-butyn-2-ol: Water   Carbon dioxide   Chemical powder   Foam

8. In the preparation of the reagents for hydration of 2-methyl-3-butyn-2-ol, why is concentrated sulfuric acid added to water rather than water added to sulfuric acid?

9. Why is the alkene added to the reaction mixture in two portions rather than all at once?

10. What is the purpose of adding potassium carbonate sesquihydrate and sodium chloride to the distillate before extracting it with dichloromethane? Why not use just potassium carbonate sesquihydrate?

11. Calculate the molar ratio of mercuric oxide and 2-methyl-3-butyn-2-ol used in this experiment. Show your calculations.

12. Is mercuric oxide the limiting reagent? Explain.

13. Why does hydration of 2-methyl-3-butyn-2-ol give the ketone, 3-hydroxy-3-methyl-2-butanone (**2**), rather than the aldehyde, 3-methyl-3-hydroxybutanal (**3**)?

14. How can **2** and **3** be differentiated chemically? by spectroscopic methods?

Pre-Lab Exercises for A. REACTION OF 1,3-BUTADIENE AND MALEIC ANHYDRIDE (Section 12.3)

NAME (print): _____ DATE: _____

INSTRUCTOR: _____ LABORATORY SECTION: _____

1. What conformation is required to enable a 1,3-diene to undergo the Diels-Alder reaction?

2. Why does 1,3-butadiene react more rapidly with maleic anhydride than with another molecule of itself?

3. Define the term "*in situ* preparation" as it applies to this experiment.

4. Why should 3-sulfolene and maleic anhydride be completely dissolved in the xylene before thermal decomposition of the sulfolene is attempted?

5. Why is a gas trap used in this experiment?

6. Why is the air being drawn through the gas trap first passed through calcium chloride?

7. What is the limiting reagent in the reaction between 1,3-butadiene and maleic anhydride? Why do you think this reagent rather than the other was made limiting?

8. Why is petroleum ether added to the solution of product in xylene?

9. What type of filtration, gravity or vacuum, should be used to isolate the crystalline product?

10. The oral LD_{50} (mg/kg) in rats of 3-sulfolene is _____ ; that of maleic anhydride is _____ .

11. The flashpoint (°C) of xylene is _____ .

12. List the target organs for xylene.

13. Underline the media, if any, that are *not* appropriate for extinguishing fires involving xylene:
    Water   Carbon dioxide   Chemical powder   Foam

Pre-Lab Exercises for B. **REACTION OF 1,3-CYCLOPENTADIENE AND MALEIC ANHYDRIDE (Section 12.3)**

NAME (print): _____ DATE: _____

INSTRUCTOR: _____ LABORATORY SECTION: _____

1. What conformation is required to enable a 1,3-diene to undergo the Diels-Alder reaction?

2. Why does 1,3-cyclopentadiene react more rapidly with maleic anhydride than with another molecule of itself?

3. Why is it important to keep the head temperature below about 45 °C when "cracking" dicyclopentadiene?

4. Why must the solution of maleic anhydride in ethyl acetate-petroleum ether be homogeneous when 1,3-cyclopentadiene is added to it?

5. Why is a 25-mL rather than a 10-mL round-bottom flask used for cracking dicyclopentadiene when only 7 mL of this substance is being cracked?

6. Why is a fractional rather than a simple distillation apparatus specified for the cracking of dicyclopentadiene?

7. What is the limiting reagent in the reaction between 1,3-cyclopentadiene and maleic anhydride? Why do you think this reagent rather than the other one was made limiting?

8. The oral LD_{50} (mg/kg) in rats of dicyclopentadiene is _____; that of maleic anhydride is _____.

9. Maleic anhydride is listed as a potentially carcinogenic compound. T_____ F_____.

10. The flashpoints (°C) of dicyclopentadiene and ethyl acetate are _____ and _____, respectively.

11. List the acute effects of exposure to vapors or mists containing dicyclopentadiene.

Pre-Lab Exercises for C. REACTION OF *cis*- AND *trans*-1,3-PENTADIENE AND MALEIC ANHYDRIDE (Section 12.3)

NAME (print): _____ DATE:_____

INSTRUCTOR: _____ LABORATORY SECTION: _____

1. What conformation is required to enable a 1,3-diene to undergo the Diels-Alder reaction?

2. Why does a 1,3-diene react more rapidly with maleic anhydride than with another molecule of itself?

3. Why is a drying tube attached to the reflux condenser for this experiment?

4. Why should a flame *not* be used for the final concentration of the reaction mixture?

5. What is wrong with adding petroleum ether to a toluene solution of the desired anhydride if this solution is at a temperature of 70 °C or so?

6. If it is assumed that both of the isomeric dienes react and that they constitute 70% of the commercial mixture used in this experiment, determine the limiting reagent in the reaction between the 1,3-pentadienes and maleic anhydride. Show your calculation.

7. What type of filtration, gravity or vacuum, should be used to isolate the product?

8. The oral LD_{50} (mg/kg) in rats of maleic anhydride is _____.

9. Maleic anhydride is listed as a potentially carcinogenic compound. T _____ F _____.

10. The flashpoints (°C) of 1,3-pentadiene and toluene are _____ and _____, respectively.

Pre-Lab Exercises for FORMATION OF SEMICARBAZONES UNDER KINETIC AND THERMODYNAMIC CONTROL (Section 13.2)

NAME (print): _____ DATE: _____

INSTRUCTOR: _____ LABORATORY SECTION: _____

1. Define:
   a. kinetic control of a reaction

   b. thermodynamic control of a reaction

2. Write an equation for the reaction between semicarbazide hydrochloride and dibasic potassium phosphate.

3. The three buffer systems described in this chapter are A: $CH_3CO_2H/CH_3CO_2^-$; B: $H_2PO_4^-/HPO_4^{2-}$; and C: H_2CO_3/HCO_3^-. Answer the following questions by using A, B, or C to refer to the proper buffer system.
   a. Which buffer system maintains the lowest pH? _____
   b. Which buffer system generally produces the fastest rate of formation of semicarbazones of aldehydes and ketones? _____
   c. Which buffer system maintains a pH in the range 6.1–6.2? _____

4. What is the purpose of cooling each reaction mixture in an ice-water bath after reaction periods of various times and at various temperatures?

5. Which experiment, C1 or C3, would you expect to be more likely to give a predominance of the product of *kinetic* control? Explain your reasoning.

6. Typical melting ranges of crystals obtained in various parts of this experiment might be, for example, 150–162 °C, 163–165 °C, and 199–201°C. Identify these melting ranges with the mixtures of semicarbazones of cyclohexanone (C) and 2-furaldehyde (F) listed below.
   a. 98% C/2%F _____
   b. 98% F/2% C _____
   c. 40% F/60% C _____

7. Why should a Thiele tube *not* be used for determining melting points in this experiment?

8. The flashpoints (°C) of cyclohexanone and 2-furaldehyde are _____ and _____, respectively.

9. The LC_{50} (ppm) for inhalation by rats of cyclohexanone is _____ during _____ hours, whereas that of 2-furaldehyde is _____ during _____ hours.

10. List the toxic fumes evolved in fires involving the semicarbazones of cyclohexanone and 2-furaldehyde.

11. 2-Furaldehyde is listed as tumorogenic compound. T _____ F _____ .

12. List the target organs for cyclohexanone.

13. List the target organs for 2-furaldehyde.

**Pre-Lab Exercises for PREPARATION OF
1-BROMOBUTANE (Section 14.4)**

NAME (print): _____ DATE: _____

INSTRUCTOR: _____ LABORATORY SECTION: _____

1. Describe the method by which hydrogen bromide is prepared for use in this experiment.

2. Determine the limiting reagent for the conversion of 1-butanol to 1-bromobutane and the theoretical yield of product.

3. Is this reaction an S_N1 or S_N2 process? How will the mechanism be confirmed experimentally?

4. Write the structures of two possible side products that might be formed in this reaction.

5. What is the purpose of the following experimental techniques, and where is each used in the preparation?

   a. heating at reflux

   b. simple distillation

   c. adding anhydrous magnesium sulfate

6. The oral LD_{50} (mg/kg) in rats of 1-butanol is _____; the intraperitoneal LD_{50} (mg/kg) in rats of 1-bromobutane is_____.

7. The flashpoints (°C) of the compounds listed in Exercise 6 are _____ and_____, respectively.

8. What toxic fumes are evolved in fires involving 1-bromobutane?

9. Water is an appropriate medium for extinguishing fires involving 1-bromobutane. T_____ F_____.

Pre-Lab Exercises for PREPARATION OF 2-CHLORO-2-METHYLBUTANE (Section 14.5)

NAME (print): _____ DATE: _____

INSTRUCTOR: _____ LABORATORY SECTION: _____

1. Determine the limiting reagent for the preparation of 2-chloro-2-methylpropane and calculate the theoretical yield of product.

2. Why is this reaction carried out at room temperature rather than at elevated temperatures?

3. Is this reaction an S_N1 or S_N2 process? Does the experiment contain any procedures for experimental verification of the mechanism?

4. Write the structures of any possible side products in the reaction.

5. After the initial reaction is carried out and the crude product is isolated, it is washed with dilute bicarbonate solution. What is the purpose of this wash, and could dilute sodium hydroxide solution be used instead of sodium bicarbonate? Explain briefly.

6. The flashpoint (°C) of 2-methyl-2-butanol is _____.

7. List the toxic fumes evolved in fires involving 2-chloro-2-methylbutane.

8. List the media appropriate for extinguishing fires involving 2-methyl-2-butanol.

**Pre-Lab Exercises for FRIEDEL-CRAFTS ALKYLATION OF
p-XYLENE WITH 1-BROMOPROPANE (Section 15.2)**

NAME (print): _____ DATE: _____

INSTRUCTOR: _____ LABORATORY SECTION: _____

1. **a.** What is the limiting reagent in this experiment?

   b. Why is this reagent chosen to be limiting?

2. **a.** What function is served by AlCl$_3$ in this reaction?

   b. Why is so little of it required?

3. Why should the p-xylene be dry?

4. Write structural formulas for the isomeric alkylation products expected from the reaction of 1-bromopropane with p-xylene.

5. Would you expect the ratio of n-propyl-p-xylene: isopropyl-p-xylene produced in this experiment to be larger or smaller than the ratio of n-propylbenzene: isopropylbenzene produced by a similar alkylation of benzene by 1-bromopropane? Explain your answer.

6. Why is the 1-bromopropane added dropwise into the reaction mixture rather than all at once?

7. Is the reaction expected to be exothermic or endothermic?

8. What is the reason for pouring the completed reaction mixture into a crushed ice-water mixture? Why water, and why ice?

9. What compounds are present in the aqueous solution that is to be discarded?

10. What hazardous combustion products are evolved in fires involving $AlCl_3$, p-xylene, and 1-bromopropane?

11. The flashpoints (°C) of p-xylene and 1-bromopropane are _____ and _____, respectively.

12. List the media appropriate for extinguishing fires involving p-xylene and 1-bromopropane.

Pre-Lab Exercises for NITRATION OF BROMO-BENZENE (Part A) (Section 15.3)

NAME (print): _____ DATE:_____

INSTRUCTOR: _____ LABORATORY SECTION: _____

1. Calculate the molar ratio of nitric acid:bromobenzene used in this experiment. (Concentrated nitric acid is 16 M.) What is the limiting reagent in this experiment?

2. Compare the ratio of reactants in this experiment (Exercise 1, above) with that in the experiment of Section 15.2 (see Pre-Lab Exercise 1 for that section). Why are they so different?

3. Why is dinitration not a significant process under the conditions used?

4. What is the function of the concentrated sulfuric acid in this experiment?

5. Why is the reaction flask shaken frequently during the addition of the bromobenzene to the mixture of acids?

6. Is the nitration reaction expected to be exothermic or endothermic?

7. What compounds are present in the aqueous solution from which the isomeric bromonitrobenzenes are filtered?

8. Why is the *p*-isomer of the product expected to be less soluble in ethanol than the *o*-isomer?

9. Circle the media, if any, that are *not* appropriate for extinguishing fires involving bromobenzene and *o*- and *p*-bromonitrobenzene: Water   Carbon dioxide   Chemical powder   Foam

10. List the toxic fumes evolved in fires involving bromobenzene and *o*- and *p*-bromonitrobenzene.

Pre-Lab Exercises for NITRATION OF BROMO-BENZENE (Part B) (Section 15.3)

NAME (print): _____ DATE:_____

INSTRUCTOR: _____ LABORATORY SECTION: _____

1. Why should the developing chamber for a TLC plate *not* be open to the atmosphere?

2. Why should a TLC plate be removed from the solvent before the solvent front reaches the top of the plate?

3. Which of the following diagrams illustrate(s) an *improper* way of spotting a TLC plate? Tell what is wrong in each such case.

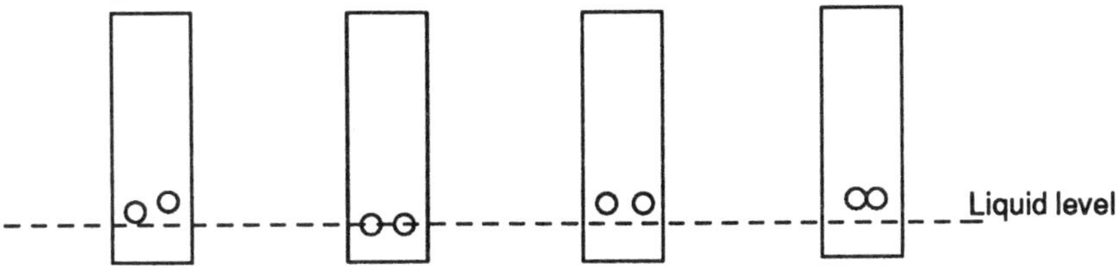

4. What difficulty may result if
   a. a chromatographic column is not placed in a vertical position?

   b. the liquid level of the eluent is allowed to drop below the top of the packing in the column?

5. Why should a minimum amount of solvent be used to introduce a mixture onto a column?

6. What would be the consequence of using a more polar eluent for both the TLC and column chromatography?

Pre-Lab Exercises for A. ACYLATION OF *m*-XYLENE WITH PHTHALIC ANHYDRIDE (Section 15.4)

NAME (print): _____ DATE: _____

INSTRUCTOR: _____ LABORATORY SECTION: _____

1. Write out a stepwise mechanism for the conversion to acid **37** of *m*-xylene (**35**) and the acylium salt **39** and specify the rate-determining step in your mechanism. Use curved arrows to symbolize flow of electrons.

2. Why does ion **35** attack at C-4 rather than C-2 of *m*-xylene?

3. Why is a gas trap a necessary part of the apparatus for this procedure and why is CaCl$_2$ rather than NaOH or KOH used in the trap?

4. Why must an ice-water bath be available at the start of the reaction?

5. Should anhydrous or technical diethyl ether be used for the extraction of the aqueous solution obtained by hydrolysis of the reaction mixture? Explain briefly.

6. In the steps of the work-up procedure requiring extraction of the ethereal solution of **37** with aqueous base and subsequent acidification of the combined extracts, approximately how much 12 N HCl (conc. HCl) will be needed to bring the pH of the extracts to 2? Show your reasoning.

7. In light of your answer in the preceding exercise, what size Erlenmeyer flask should be used to contain the basic extracts?

8. List the target organs for toxic effects of *m*-xylene.

9. What toxic fumes are evolved in fires involving $AlCl_3$?

10. The values of the oral LD_{50} (mg/kg) in rats of $AlCl_3$, *m*-xylene, and phthalic anhydride are _____ , _____ , and _____ , respectively.

Pre-Lab Exercises for B. **PREPARATION OF 1,3-DIMETHYL-ANTHRAQUINONE (Section 15.4)**

NAME (print): _____ DATE: _____

INSTRUCTOR: _____ LABORATORY SECTION: _____

1. Write out a stepwise mechanism for the conversion of **37** to **38** in the presence of sulfuric acid, and specify the rate-determining step in your mechanism. Use curved arrows to symbolize flow of electrons.

2. Determine the limiting reagent in this reaction. Show your work.

3. Why should the reaction mixture containing **38** be allowed to cool to 100 °C before water is added to it?

4. No MSDS information is currently available for the dimethylanthraquinone **38**. In lieu of this, look up the corresponding data for anthraquinone itself and summarize it below.

5. List the toxic fumes evolved in fires involving sulfuric acid.

Pre-Lab Exercises for C. REDUCTION OF 1,3-DIMETHYL ANTHRAQUINONE (Section 15.4)

NAME (print): _____ DATE: _____

INSTRUCTOR: _____ LABORATORY SECTION: _____

1. Write the balanced equation for the reduction of **38** to **43** with $SnCl_2$ and acetic acid.

2. Write a stepwise mechanism for the formation of **41** from **43** and specify the rate-determining step in your mechanism. Use curved arrows to symbolize flow of electrons.

3. What is the limiting reagent in this reaction?

4. What is the reducing agent in this reaction and in what oxidation state does it exist at the beginning of the reaction? at the end?

5. The flashpoint (°C) of acetic acid is _____.

6. Do you expect a heterogeneous or homogeneous mixture at the beginning of the reaction?

7. List the toxic materials evolved by fires involving stannous chloride and give the extinguishing medium appropriate for use in such a fire.

Pre-Lab Exercises for D. PREPARATION OF 1,3-DIMETHYL ANTHRACENE (Section 15.4)

NAME (print): _____ DATE: _____

INSTRUCTOR: _____ LABORATORY SECTION: _____

1. Write the balanced equation for the reduction of the dimethylanthrones to 1,3-dimethyl-anthracene.

2. What is the limiting reagent in the reduction reaction? Show your work.

3. Write the balanced equation for the reaction of sodium borohydride with aqueous acid.

4. Compute the amount of 6 M HCl required to neutralize the reaction mixture. Show your work.

5. Why is it unnecessary to protect the reaction mixture from atmospheric moisture?

6. No MSDS information is currently available for 1,3-dimethylanthracene. In lieu of this, look up the corresponding data for anthracene and summarize it below.

7. Underline the media appropriate for extinguishing fires involving sodium borohydride:
   Water   Carbon dioxide   Chemical powder   Foam

8. What gas, if any, is evolved in the reaction of Exercise 3?

9. The flashpoints (°C) of isopropyl alcohol, methanol and toluene are _____, _____, and _____, respectively.

Pre-Lab Exercises for RELATIVE RATES OF ELECTROPHILIC AROMATIC SUBSTITUTION (Section 15.5)

NAME (print): _____ DATE:_____

INSTRUCTOR: _____ LABORATORY SECTION: _____

1. Why is 15 M acetic acid an appropriate solvent in which to perform rate studies of electrophilic brominations?

2. Show that 15 M acetic acid is approximately 90% acetic acid and 10% water by weight.

3. Write structural formulas of the substrates used in this experiment.

4. Why is benzene not used as a substrate in this experiment?

PL. 71

5. What criterion is used for measuring the rates of bromination of the aromatic compounds?

6. Why is it important to maintain a constant temperature in this experiment?

7. Why might it be necessary to conduct this experiment at more than one temperature?

8. The flashpoints (°C) of anisole, diphenyl ether, and phenol are _____, _____, and _____, respectively.

9. Underline the media, if any, that are *not* appropriate for extinguishing fires involving the compounds listed in Exercise 8:   Water   Carbon dioxide   Chemical powder   Foam

**Pre-Lab Exercises for PREPARATION OF
2-METHYLPROPANAL (Section 16.2)**

NAME (print): _____ DATE:_____

INSTRUCTOR: _____ LABORATORY SECTION:_____

1. Write the balanced equation for the oxidation of 2-methyl-1-propanol to isobutyl 2-methylpropanoate by chromic acid prepared from potassium dichromate and aqueous sulfuric acid. Show all work.

2. Which is normally the rate-determining step in the oxidation of alcohols, formation of a chromate ester or elimination of this ester?

3. What is the limiting reagent in this reaction? Show your work.

4. What is the consequence of failing to distil the aldehyde from the oxidizing solution as quickly as possible?

5. Why is concentrated sulfuric acid added *to* water rather than the reverse?

6. Why does the initially red solution of chromic acid turn green as the oxidation progresses?

7. Why is a Hempel column interposed between the Claisen adapter and the stillhead during the oxidation stage of this reaction?

8. What is the purpose of washing the steam distillate with sodium carbonate?

9. The flashpoints (°C) of 2-methyl-1-propanol-2-ol and 2-methylpropanal are _____ and _____, respectively.

10. Potassium dichromate is listed as a potentially carcinogenic compound. T_____ F_____.

11. 2,4-Dinitrophenylhydrazine has been show to have mutagenic effects. T_____ F_____.

12. List the toxic products evolved in fires involving 2,4-dinitrophenylhydrazine.

Pre-Lab Exercises for PREPARATION OF CYCLOHEXANONE (Section 16.2)

NAME (print): _____ DATE: _____

INSTRUCTOR: _____ LABORATORY SECTION: _____

1. Write the balanced equation for the oxidation of cyclohexanol to cyclohexanone by chromic acid prepared from potassium dichromate and sulfuric acid. Show all work.

2. Which is normally the rate-determining step in the oxidation of alcohols, formation of a chromate ester or elimination of this ester?

3. What is the limiting reagent in this reaction? Show your work.

4. Why do you think this reagent rather than any other was made limiting?

5. Why does the initially red solution of chromic acid change color as oxidation proceeds?

6. Why is concentrated sulfuric acid added *to* an aqueous solution of potassium dichromate rather than the reverse for the preparation of chromic acid?

7. Why is a fractional rather than a simple distillation apparatus used for the initial distillation of the reaction mixture?

8. 2,4-Dinitrophenylhydrazine has been shown to have mutagenic effects. T _____ F _____.

9. The flashpoints (°C) of cyclohexanol and cyclohexanone are _____ and _____, respectively.

10. Potassium dichromate is listed as a potentially carcinogenic compound. T _____ F _____.

11. List the toxic products evolved in fires involving 2,4-dinitrophenylhydrazine.

Pre-Lab Exercises for OXIDATION OF CYCLOHEXANONE TO HEXANEDIOIC ACID (Section 16.2)

NAME (print): _____ DATE: _____

INSTRUCTOR: _____ LABORATORY SECTION: _____

1. Write the balanced equation for the oxidation of cyclohexanone to dipotassium hexanedioate by basic potassium permanganate. Show all work.

2. What transformation must the ketone undergo before it is attacked by permanganate ion?

3. What is the limiting reagent in this reaction? Show all work.

4. Why do you think this reagent rather than the other was made limiting?

5. Why is it important to destroy all of the potassium permanganate before work-up of the reaction mixture?

6. Write the reaction that occurs between sodium bisulfite and potassium permanganate in aqueous solution.

7. Why does the initially purple solution of potassium permanganate change color as the oxidation progresses?

8. Why would this oxidative method for the preparation of carboxylic acids be unsuitable for the synthesis of pentanoic acid from 4-octanone or 4-nonanone?

9. The LC_{50} (ppm) for inhalation by rats of cyclohexanone is _____ during _____ hours, and the target organs for its action is (are):

10. List the target organs for toxicological action of potassium permanganate and manganese dioxide.

Pre-Lab Exercises for OXIDATION OF CYCLODODECANOL TO CYCLODODECANONE (Section 16.2)

NAME (print): _____ DATE: _____

INSTRUCTOR: _____ LABORATORY SECTION: _____

1. What is the oxidizing agent in this experiment and to what is it reduced in the course of the reaction?

2. What qualitative test is performed to determine whether sufficient oxidizing agent has been used in this procedure?

3. Write an equation that expresses the chemical reaction that forms the basis of the test of Exercise 2.

4. What purpose is served by washing the ethereal extract containing the product with saturated sodium bicarbonate?

5. The flashpoints (°C) of acetone and acetic acid are _____, and _____, respectively.

6. An MSDS is not presently available on either cyclododecanol or cyclododecanone. In lieu of this information, provide the relevant toxicological information for the analogous compounds, cyclohexanol and cyclohexanone.

7. Underline the media, if any, that are *not* appropriate for extinguishing fires involving cyclododecanol and cyclododecanone (assume these are the same as for cyclohexanol and cyclohexanone, respectively): Water Carbon dioxide Chemical powder Foam

Pre-Lab Exercises for BASE CATALYZED OXIDATION-REDUCTION OF ALDEHYDES: THE CANNIZZARO REACTION (Section 16.3)

NAME (print): _____ DATE: _____

INSTRUCTOR: _____ LABORATORY SECTION: _____

1. Write the balanced equation for the oxidation-reduction of benzaldehyde to benzyl alcohol and potassium benzoate by reaction with potassium hydroxide. Show all work.

2. Why is the Cannizzaro reaction limited to aldehydes having no α-hydrogen atoms?

3. What is the limiting reagent in this reaction? Show how you arrived at this conclusion.

4. Why do ketones having no α-hydrogen atoms not undergo the Cannizzaro reaction?

5. What is an emulsion? Why is it desirable to have an emulsion rather than two phases during the reaction between benzaldehyde and aqueous base?

6. What is the solid that is formed after benzaldehyde has been allowed to react with aqueous potassium hydroxide? Why does it dissolve in water?

7. Why should a water-cooled condenser *not* be used in the isolation of benzyl alcohol by distillation, once the diethyl ether has been removed?

8. The oral LD_{50} (mg/kg) in rats of benzaldehyde is _____; that of benzyl alcohol is _____.

Pre-Lab Exercises for CARBOXYLIC ACIDS
(Section 16.4)

NAME (print): _____ DATE: _____

INSTRUCTOR: _____ LABORATORY SECTION: _____

1. By calculating the change in oxidation number at the benzylic and the methyl carbon atoms in the conversion of ethylbenzene to benzoate and carbonate ions, and by reference to Equation (16.28), determine how many moles of permanganate are required to oxidize one mole of ethylbenzene to the aforementioned products. Show all work.

2. What is the limiting reagent in this reaction? Show how you arrived at this conclusion?

3. Why is it necessary to destroy residual potassium permanganate *prior* to work-up of the reaction mixture?

4. Write the equation for the reaction between sodium bisulfite and potassium permanganate.

5. Why is filter-aid added to the reaction mixture *prior* to its vacuum filtration?

6. Why is hydrochloric acid added to the aqueous residue obtained by simple distillation of the filtrate? Why might concentrated rather than dilute acid be preferred for this step?

7. The LDL_0 of benzoic acid in humans is _____; the oral LD_{50} (mg/kg) in rats of ethylbenzene is _____.

8. List the target organs for toxicological action of potassium permanganate and manganese dioxide.

**Pre-Lab Exercises for HYDROGENATION OF 4-CYCLOHEXENE-
cis-1,2-DICARBOXYLIC ACID   (Section 17.2)**

NAME (print): _____ DATE:_____

INSTRUCTOR: _____ LABORATORY SECTION: _____

1. What is meant by the term *catalytic hydrogenation*?

2. What catalyst is used in this preparation, and how is it prepared?

3. Indicate how hydrogen gas is generated in this preparation; write an equation to show this reaction.

4. What is the stereochemistry of catalytic hydrogenation? Cite an example that shows this.

5. Does the catalytic hydrogenation of 4-cyclohexene-*cis*-1,2-dicarboxylic acid show the stereochemistry of hydrogenation? Explain.

6. What is the solvent for this preparation?

7. Could water have been used as the solvent instead of the one that is used? Explain.

8. Does addition of concentrated HCl to an aqueous solution of the reduction product of this reaction increase or decrease the water solubility of the product? Why?

9. What hazardous mineral acid is produced by decomposition of chloroplatinic acid?

10. Underline the media appropriate for extinguishing fires involving sodium borohydride:
    Water   Carbon dioxide   Chemical powder   Foam

11. Write the balanced equation for reaction of sodium borohydride with concentrated HCl.

12. What MSDS toxicological data, if any, are available for 4-cyclohexene-1,2-dicarboxylic acid?

**Pre-Lab Exercises for FORMATION AND REDUCTION OF
N-CINNAMYLIDENE-m-NITROANILINE
(Section 17.2)**

NAME (print): _____ DATE:_____

INSTRUCTOR: _____ LABORATORY SECTION: _____

1. Draw the structure of an imine, and give the equation for its conversion into an amine as done in this experiment.

2. What metal hydride reducing agent is used in this experiment, and why is it more convenient to do so in preference to using hydrogen gas?

3. This experiment is an example of *reductive amination*. Define this term, and indicate the advantage of using the experimental procedure described rather than hydrogen gas. What product would be formed if catalytic hydrogenation were used?

4. Draw the structure of the imine intermediate that is formed, and describe what experimental technique is used to maximize the yield of this compound.

5. Why is it unnecessary to isolate and purify the intermediate imine?

6. Determine the limiting reagent in this preparation and compute the theoretical yield. Show all work.

7. List the toxic fumes evolved in a fire involving cinnamaldehyde, *m*-nitroaniline, and cyclohexane.

8. The oral LD_{50} (mg/kg) in rats of *m*-nitroaniline is _____.

9. The flashpoints (°C) of cyclohexane and cinnamaldehyde are _____, and _____, respectively.

10. Underline the media appropriate for extinguishing fires involving sodium borohydride: Water   Carbon dioxide   Chemical powder   Foam

11. Write an equation for reaction of sodium borohydride and methanol that yields hydrogen.

Pre-Lab Exercises for REDUCTION OF FLUORENONE (Section 17.4)

NAME (print): _____ DATE: _____

INSTRUCTOR: _____ LABORATORY SECTION: _____

1. Determine whether sodium borohydride is the limiting reagent in this experiment. Show your work.

2. Why is it important to avoid exposing sodium borohydride to moisture? Write the reaction that occurs when sodium borohydride and water are mixed.

3. What color change should occur as the reduction of fluorenone proceeds? Provide a reason for this change.

4. In the experiment, the reducing agent is allowed to react with fluorenone, and after the reaction is complete, sulfuric acid is added. What is the purpose of this addition? Why is it important to dissolve all of the solids completely?

5. Describe the advantages of using a metal hydride reduction reaction rather than catalytic hydrogenation in this procedure.

6. Write an equation for the reaction that might occur during the recrystallization of fluorenol from methanol if all of the sulfuric acid has not been removed by washing.

7. What toxicological data are available on the MSDS for fluorenol? for fluorenone?

8. Underline the media, if any, that are *not* appropriate for extinguishing fires involving a combination of fluorenone, fluorenol, methanol, and sodium borohydride: Water Carbon dioxide Chemical powder Foam

Pre-lab Exercises for ENZYMATIC REDUCTION OF METHYL ACETOACETATE (Section 17.5)

NAME (print): _____ DATE: _____

INSTRUCTOR: _____ LABORATORY SECTION: _____

1. Why should the Erlenmeyer flask in which the fermentation is conducted not be stoppered tightly?

2. What is the role of the Na_2HPO_4?

3. What is the advantage of using dichloromethane as the extraction solvent rather than diethyl ether?

4. Why use filter aid for filtering the reaction mixture?

5. Why is it important *not* to shake the separatory funnel vigorously during the extraction of the product with dichloromethane?

6. Although sucrose is added to the reaction, it is not necessary. Propose a reason for using sucrose, and account for the fact that it is not absolutely required for the reduction.

7. The oral LD$_{50}$ (mg/kg) in rats of sucrose is _____, that of methyl acetoacetate is _____.

8. What toxic fumes are evolved upon combustion of dichloromethane?

Pre-lab Exercises for DETERMINING OPTICAL PURITY OF METHYL 3-HYDROXYBUTANOATE (Section 17.6)

NAME (print): _____ DATE: _____

INSTRUCTOR: _____ LABORATORY SECTION: _____

1. Why is it important to use a dry NMR tube for measuring the optical purity of a sample using a chiral shift reagent?

2. Why should the chiral shift reagent be stored in a desiccator?

3. What is the effect of having undissolved solid in the NMR sample?

4. Why should the sample be allowed to stand for 20 min after dissolution of the shift reagent?

5. What is the estimated accuracy of determining the enantiomeric excess by the NMR method?

6. When determining the optical purity of an alcohol using chiral shift reagents, it is important to perform the experiment for samples of both the racemic alcohol and enantiomerically enriched alcohol. Explain.

7. The ratio of the two enantiomers is determined by comparing the peak areas for the methyl peaks on the ester group. Why is this better than trying to compare the peak areas for the methine hydrogen α to the hydroxyl group?

8. List the toxic products evolved in fires involving chloroform.

9. Chloroform is listed as a potential carcinogen. T _____ F _____

Pre-Lab Exercises for WADSWORTH-EMMONS MODIFICATION
OF THE WITTIG REACTION (Section 18.1)

NAME (print): _____ DATE: _____

INSTRUCTOR: _____ LABORATORY SECTION: _____

1. What is the limiting reagent in this experiment? Show your work.

2. What happens to the ethyl chloride that is produced in this reaction?

3. Why is triethyl phosphite used in this experiment rather than triphenylphosphine?

4. Why should sodium methoxide not be exposed to the atmosphere for more than a few minutes?

5. Why is *N,N*-dimethylformamide a better solvent for the reaction of sodium methoxide with the phosphonate ester, **12**, than methanol, for example?

6. Suppose the aldehyde used in a Wittig reaction is contaminated with the corresponding carboxylic acid. What complication would this cause?

7. Is the reaction of the sodium salt **13** of the phosphonate ester with benzaldehyde exothermic or endothermic? Give the basis for your answer.

8. How is the product of this experiment freed from residual benzaldehyde?

9. The oral LD_{50} (mg/kg) in rats of triethyl phosphite is _____; those of benzyl chloride, benzaldehyde, and cinnamaldehyde are _____, _____, and _____, respectively.

10. N,N-Dimethylformamide is listed as a potential carcinogen. T_____ F_____.

11. The flashpoint (°C) of triethyl phosphite is _____.

Pre-Lab Exercises for ALDOL CONDENSATIONS   (Section 18.2.1)

NAME (print): _____ DATE: _____
INSTRUCTOR: _____ LABORATORY SECTION: _____

1. Either *p*-anisaldehyde and acetophenone or 2-furaldehyde and *m*-nitroacetophenone are used in equimolar amounts in this experiment. Suggest a complication that might result if two moles of ketone per mole of aldehyde were used.

2. The amount of sodium hydroxide used to promote the condensation reaction is much less than an equimolar amount. What complication might result if a much larger amount of sodium hydroxide were used?

3. Why is a larger volume of 95% ethanol used as a solvent for the reaction of 2-furaldehyde and *m*-nitroacetophenone than for the reaction of anisaldehyde and acetophenone? (*Hint*: Consider the effect of nitro groups on the physical properties of organic compounds.)

4. Why is the condensation (dehydrated) product rather than the aldol addition (hydrated) product obtained in this experiment?

5. Why does the enolate ion of an aromatic ketone react faster with an aldehyde group, producing a crossed-aldol reaction, than with the carbonyl group of another molecule of ketone?

6. The oral LD_{50} (mg/kg) in rats of acetophenone is _____; the corresponding value for p-anisaldehyde is _____.

7. The oral LD_{50} (mg/kg) in rats of *m*-nitroacetophenone is _____; the corresponding value for 2-furaldehyde is _____.

Pre-Lab Exercises for REACTIONS OF α,β-UNSATURATED KETONES (Section 18.2.2)

NAME (print): _____ DATE: _____

INSTRUCTOR: _____ LABORATORY SECTION: _____

1. Write a balanced equation for the reaction of *p*-anisalacetophenone with semicarbazide.

2. Write a balanced equation for the reaction of *p*-anisalacetophenone with hydroxylamine.

3. Write configurational formulas for the *erythro* and *threo* isomers of anisalacetophenone dibromide.

4. The oral LD_{50} (mg/kg) in rats of hydroxylamine hydrochloride is _____; the corresponding value for semicarbazide hydrochloride is _____.

5. List the toxic fumes evolved in fires involving hydroxylamine hydrochloride and semi-carbazide hydrochloride.

6. What is the vapor pressure of bromine at 20 °C?

7. Why should inhalation of the vapors of bromine be avoided?

Pre-Lab Exercises for ALKYLATION OF DIMETHYL MALONATE (Section 18.2.3)

NAME (print): _____ DATE:_____

INSTRUCTOR: _____ LABORATORY SECTION: _____

1. What is the molar ratio of 1-bromohexane to dimethyl malonate used in this experiment? Show your work.

2. What effect would using a molar ratio of 1-bromohexane to dimethyl malonate of much greater than 1:1 have on the experiment?

3. Why would KOH be an inappropriate base for use in this reaction?

4. Why would $NaHCO_3$ be an inappropriate base for use in this reaction?

5. Explain why water is an inappropriate medium for extinguishing fires involving metallic sodium.

6. How is mineral oil removed from sodium metal in this procedure?

7. Why should apparatus and reagents for this experiment be dry?

8. Why add the 1-bromohexane *to* the dimethyl sodiomalonate rather than the reverse?

9. The flashpoints (°C) of methanol and dimethyl malonate are _____ and _____, respectively.

10. The oral LD_{50} (mg/kg) in rats of dimethyl malonate is _____ with respect to inhalation by rats, the LC_{50} of this compound is _____ (give value and its associated units).

11. List the toxic fumes evolved in fires involving 1-bromohexane.

Pre-Lab Exercises for PREPARATION OF 4,4-DIMETHYL-2-CYCLOHEXEN-1-ONE (Section 18.3)

NAME (print): _____ DATE: _____

INSTRUCTOR: _____ LABORATORY SECTION: _____

1. Why is an *intramolecular* rather than an *inter*molecular aldol condensation favored with 2,2-dimethylhexanal-5-one (51)?

2. What is the limiting reagent in the reaction to form the Michael addition product 51? Show your work.

3. Why do you think the carbonyl-containing compound identified in Exercise 2 rather than the other was made limiting?

4. Why must water be removed from the reaction mixture during the course of the reaction?

5. 1-Propanol boils at only a slightly lower temperature than toluene, and all the reagents used in this experiment are soluble in it. Why is it an inappropriate solvent to use in this procedure?

6. This reaction occurs much faster if a sulfonic acid rather than a carboxylic acid is used as the catalyst. Why might this be?

7. What is the theoretical volume of water that would be obtained in the reaction? Show your work. How might you account for the production of *greater* than a theoretical volume in the experiment?

8. Why should water *not* be flowing through the condenser at the time when the final product is distilling, if the distillation is performed at atmospheric pressure?

9. The flashpoints (°C) of toluene, 3-buten-2-one, and 2-methylpropanal are _____, _____, and _____, respectively.

10. The oral LD_{50} (mg/kg) in rats of 3-buten-2-one is _____; the corresponding value for 2-methylpropanal is _____.

11. 3-Buten-2-one is listed as a potential carcinogen. T _____ F _____.

Pre-Lab Exercises for PREPARATION OF THE GRIGNARD REAGENT (Section 19.2, Part A)

NAME (print): _____ DATE: _____

INSTRUCTOR: _____ LABORATORY SECTION: _____

1. What is the limiting reagent in this reaction? Show your work.

2. Why are ethereal solvents important to the success of preparing the Grignard reagent?

3. Why must the reagents, solvents, and apparatus used for preparing the Grignard reagent be dry?

4. Why is it necessary to have an ice-water bath available during the preparation of the Grignard reagent?

5. What signs should you look for in determining whether the reaction initiated?

6. Why should the alkyl halide from which the Grignard reagent is to be made *not* be added all at once to the reaction flask?

7. Why is anhydrous diethyl ether added to the magnesium in two portions, one at the beginning of the reaction and the second after formation of the Grignard reagent has started?

8. The flashpoints (°C) of diethyl ether, bromobenzene, and 1-bromobutane are _____, _____, and _____, respectively.

9. 1,2-Dibromoethane is *not* listed as a potential (underline)   Carcinogen   Hallucinogen   Tumorigen

10. List the toxic fumes evolved in fires involving diethyl ether and either bromobenzene or 1-bromobutane.

**Pre-Lab Exercises for PREPARATION OF TRIPHENYL-
METHANOL (Section 19.2, Part B)**

NAME (print): _____ DATE: _____

INSTRUCTOR: _____ LABORATORY SECTION: _____

1. What is the limiting reagent in this reaction? Show your work.

2. Why do you think the particular reagent specified in Exercise 1 was made limiting?

3. Why would the presence of methanol in the methyl benzoate lower the yield of triphenyl-methanol?

4. Why must the ester and the diethyl ether in which it is dissolved be anhydrous?

5. Why is the solution of ester added to the Grignard reagent in a dropwise fashion rather than all at once?

6. What is wrong with storing ethereal solutions in the laboratory bench from one period to the next?

7. Why is technical rather than anhydrous diethyl ether used in the work-up procedure?

8. Why is saturated aqueous sodium chloride rather than just water used to remove residual sulfuric acid from the organic solution during the work-up procedure?

9. How is unchanged ester removed from the desired product by the work-up procedure used?

10. What happens to any Grignard reagent that remains in the reaction mixture after addition of the ester?

11. The flashpoint (°C) of methyl benzoate is _____; its oral LD_{50} (mg/kg) in rats is _____.

Pre-Lab Exercises for PREPARATION OF BENZOIC ACID (Section 19.2, Part C)

NAME (print): _____ DATE: _____

INSTRUCTOR: _____ LABORATORY SECTION: _____

1. What is the limiting reagent in this procedure? Show your work.

2. Why do you think the particular reagent specified in Exercise 1 was made limiting?

3. What would the effect on yield be of allowing the crushed Dry Ice to be exposed to the atmosphere for long periods of time before use?

4. Why does pressure develop in the separatory funnel when the ethereal solution of benzoic acid is being extracted with aqueous sodium hydroxide?

5. Why is technical rather than anhydrous diethyl ether used in the work-up of the reaction mixture?

6. How are possible by-products such as biphenyl and benzophenone removed from the desired benzoic acid by the work-up procedure?

7. What is wrong with storing ethereal solutions in the laboratory bench from one period to the next?

8. The flashpoint (°C) of cyclohexane is _____.

9. The oral LDL_o (mg/kg) in humans of benzoic acid is _____.

Pre-Lab Exercises for PREPARATION OF 2-METHYL-3-HEPTANOL (Section 19.2, Part D)

NAME (print): _____ DATE: _____

INSTRUCTOR: _____ LABORATORY SECTION: _____

1. What is the limiting reagent in this procedure? Show your work.

2. Why do you think the particular reagent specified in Exercise 1 was made limiting?

3. Why is it recommended that 2-methylpropanal be freshly distilled?

4. Why must the aldehyde and the diethyl ether in which it is dissolved be anhydrous?

5. Why is the solution of aldehyde added to the Grignard reagent in a dropwise fashion rather than added all at once?

6. What is wrong with storing ethereal solutions in the laboratory bench from one period to the next?

7. Why is technical rather than anhydrous diethyl ether used in the work-up procedure?

8. Why is hydrolysis of the reaction mixture performed with aqueous sulfuric acid that is cold rather than at room temperature?

9. How is unchanged aldehyde removed from the desired product by the work-up procedure used?

10. What happens to any Grignard reagent that remains in the reaction mixture after addition of the aldehyde?

11. The flashpoint (°C) of 2-methylpropanal is _____; its oral LD_{50} (mg/kg) in rats is _____.

12. MSDS information is presently not available for 2-methyl-3-heptanol. In lieu of such data, provide the corresponding information for the analogous compound, 1-heptanol.

Pre-Lab Exercises for PREPARATION OF ANILINE (Section 20.2, Part A)

NAME (print): _____ DATE:_____

INSTRUCTOR: _____ LABORATORY SECTION: _____

1. What is the limiting reagent in this reaction? Show your work.

2. Why is the reduction of nitrobenzene more efficient with tin powder that is free of surface oxides?

3. Specify the gas that is evolved upon addition of concentrated hydrochloric acid to tin powder, and write a balanced equation for its formation.

4. Why is the concentrated hydrochloric acid added in portions rather than all at once?

5. How is aniline separated from residual nitrobenzene in this procedure?

6. Why is sodium chloride added to the steam distillate after it has been made basic but before extraction with diethyl ether?

7. Why is the reaction mixture cooled after being saturated with salt but before extraction with diethyl ether?

8. What is the color of pure aniline? Why are undistilled samples of it often seen to be brown in color?

9. The oral LD_{50} (mg/kg) in rats of nitrobenzene and aniline are _____, and _____, respectively.

10. Aniline is listed as a potential carcinogen. T _____ F _____.

Pre-Lab Exercises for PREPARATION OF ACETANILIDE (Section 20.2, Part B)

NAME (print): _____ DATE:_____

INSTRUCTOR: _____ LABORATORY SECTION: _____

1. What is the limiting reagent in this reaction? Show your work.

2. How is unchanged aniline separated from the desired acetanilide in this procedure?

3. Suppose you had to make up the specified aqueous solution of sodium acetate *after* you added the acetic anhydride to the aqueous solution of aniline. What would be the consequence of this? Why?

4. What is the reason that aqueous hydrochloric acid is combined with aniline in this procedure?

5. Why is concentrated hydrochloric acid added to water rather than the reverse in the preparation of dilute acid used in this procedure?

6. What is the role of the sodium acetate in this reaction?

7. Aniline is listed as a potential carcinogen. T _____ F _____ .

8. List the toxic fumes evolved in fires involving aniline, acetic anhydride, and acetanilide.

9. The oral LD_{50} (mg/kg) in rats of aniline and acetic anhydride are _____ and _____, respectively.

**Pre-Lab Exercises for PREPARATION OF *p*-ACETAMIDOBENZENE-
SULFONYL CHLORIDE (Section 20.2, Part C)**

NAME (print): _____ DATE:_____

INSTRUCTOR: _____ LABORATORY SECTION: _____

1. What is the limiting reagent in this reaction? Show your work.

2. Why is it important that the acetanilide used in this experiment be dry?

3. What color does the reaction mixture turn if the chlorosulfonation is proceeding normally?

4. Why is ice water rather than warm water used for hydrolysis of the reaction mixture?

5. Given the work-up procedure that is used, could residual acetanilide contaminate the desired product in its crude state? Explain.

6. Why should the *p*-acetamidobenzenesulfonyl chloride be combined with ammonia immediately rather than in the next laboratory period?

7. List the toxic fumes evolved in fires involving chlorosulfonic acid and acetanilide.

8. Why is water *not* an appropriate medium for extinguishing fires involving chlorosulfonic acid?

9. An MSDS is currently not available for *p*-acetamidobenzenesulfonyl chloride. In lieu of this information, provide the corresponding information for the analogous compound benzenesulfonyl chloride.

Pre-Lab Exercises for PREPARATION OF SULFANILAMIDE (Section 20.2, Part D)

NAME (print): _____ DATE: _____

INSTRUCTOR: _____ LABORATORY SECTION: _____

1. What is the limiting reagent in this reaction? Show your work.

2. Why does reaction of *p*-acetamidobenzenesulfonyl chloride with ammonium hydroxide give the sulfonamide rather than the sulfonic acid?

3. Why is 6 *M* sulfuric acid added after reaction of the sulfonyl chloride with ammonium hydroxide?

4. Why might there be a preference for the use of solid sodium carbonate instead of solid sodium hydroxide for basifying the acidic hydrolysis solution obtained in this experiment?

5. The oral LD_{50} (mg/kg) in rats of sulfanilamide is _____.

6. List the toxic fumes evolved in fires involving sulfanilamide.

7. Sulfanilamide is listed as a potential tumorigen. T_____ F_____.

Pre-Lab Exercises for PREPARATION OF POLYSTYRENE (Section 21.2)

NAME (print): _____ DATE: _____
INSTRUCTOR: _____ LABORATORY SECTION: _____

1. Write equations illustrating a mechanism for cationic chain-reaction polymerization of 2-methylpropene (isobutylene).

2. Write equations illustrating a mechanism for anionic chain-reaction polymerization of propenenitrile (acrylonitrile).

3. Write an equation for the reaction of *tert*-butylcatechol with sodium hydroxide solution that is responsible for removing the inhibitor from commercial styrene and explain how the extraction accomplishes the desired separation.

4. Write formulas for the products of the *disproportionation* reaction between two styryl radicals, $C_6H_5 CHCH_3$.

5. What determines whether or not decantation rather than filtration may be used to separate a solid from a liquid?

6. Styrene and *tert*-butyl peroxybenzoate are both listed as potential tumorigenics. T _____ F _____.

7. The LCL_{50} (ppm) for inhalation of styrene by humans is _____ during _____ min.

8. The flashpoints (°C) of styrene, *tert*-butyl peroxybenzoate, and xylene are _____, _____, and _____, respectively.

Pre-Lab Exercises for PREPARATION OF
NYLON-6,10 (Section 21.3)

NAME (print): _____ DATE: _____

INSTRUCTOR: _____ LABORATORY SECTION: _____

1. Why is decanedioyl chloride rather than decanedoic acid used in this experiment?

2. Explain how reaction between decanedioyl chloride in dichloromethane solution and 1,6-hexanediamine in aqueous solution occurs although the two immiscible solutions are not mixed.

3. Why should the tip of the separatory funnel containing the 1,6-hexanediamine solution be placed no more than 1 cm above the surface of the dichloromethane solution when making the addition?

4. Why is the aqueous solution added to the dichloromethane solution, rather than *vice versa*?

5. Note that decanedioyl chloride and 1,6-hexanediamine are used in equimolar amounts, whereas in many experiments one reactant is used in considerable molar excess. Why is the equimolar ratio used here?

6. Why is the formic acid solution of the polymer evaporated in the hood rather than at the laboratory bench?

7. The flashpoint (°C) of decanedioyl dichloride is _____.

8. List the toxic fumes that are evolved in fires involving dichloromethane, 1,6-hexane-diamine and decanedioyl chloride.

9. The oral LD_{50} (mg/kg) in rats of 1,6-hexanediamine is _____, whereas that of formic acid is _____.

Pre-Lab Exercises for HYDROLYSIS OF SUCROSE (Section 22.2)

NAME (print): _____ DATE: _____

INSTRUCTOR: _____ LABORATORY SECTION: _____

1. Why is the mixture of sugars derived from hydrolysis of sucrose commonly referred to as "invert" sugar?

2. What effect would the presence of air bubbles in the polarimeter tube have on the value of the measured rotation?

3. What effect would incomplete transfer of the hydrolysis mixture from the reaction flask to the volumetric flask have on the value of the measured rotation?

4. The oral LD_{50} (mg/kg) in rats of sucrose is _____.

5. Sucrose is a reducing sugar. T _____ F _____

6. What is the role of hydrochloric acid in the hydrolysis of sucrose?

Pre-Lab Exercises for CLASSIFICATION TESTS FOR CARBOHYDRATES (Section 22.3.1)

NAME (print): _____ DATE: _____

INSTRUCTOR: _____ LABORATORY SECTION: _____

1. Tollens' and Benedict's tests depend on the ability of carbohydrates to oxidize the test reagent. T _____ F _____

2. Explain why nonreducing sugars give a positive Barfoed's test more slowly than do reducing sugars.

3. What is likely to happen if Tollens' test is performed in a test tube that is not clean?

4. What do you expect to see in the case of a positive Benedict's test?

5. What do you expect to see in the case of a positive Barfoed's test?

6. Why are both glucose and sucrose recommended as carbohydrates to test for purposes of comparison when performing Benedict's test on an unknown?

7. Why is it important not to heat the test solution for longer than 5 minutes when performing Barfoed's test?

8. What purpose is served by citrate ion in the stock solution of Benedict's reagent?

Pre-Lab Exercises for FORMATION OF OSAZONES (Section 22.3.2)

NAME (print): _____ DATE: _____
INSTRUCTOR: _____ LABORATORY SECTION: _____

1. What is the function of saturated aqueous sodium bisulfite solution in this procedure?

2. Would you expect osazone formation to occur faster or slower in strongly acidic media as compared to weakly acidic media? Explain your answer.

3. Why does reaction of phenylhydrazine with aldoses and ketoses stop after the first two carbon atoms in the chain have been converted to hydrazone functions?

4. Why is sodium acetate required if phenylhydrazine hydrochloride is used as the source of phenylhydrazine?

5. Why would D-glucose and D-mannose be expected to form the same osazone upon reaction with phenylhydrazine?

6. Phenylhydrazine and its hydrochloride are both listed as a potentially carcinogenic compounds. T_____ F_____.

7. List the toxic fumes evolved upon combustion of phenylhydrazine and its hydrochloride.

8. Underline the media that are appropriate for extinguishing fires involving glucose, fructose, and phenylhydrazine hydrochloride: Water Carbon dioxide Chemical powder Foam